Exploring with

Power Polygons

Mary Kay Tornrose-Dyer

Cindy Stephens

Cuisenaire Company of America, Inc.
White Plains, New York

Managing editor: Doris Hirschhorn
Development editor: Linda Carbone
Design and Production: Woodshed Productions

ISBN 0-938587-79-X

1 2 3 4 5 00 99 98 97 96

TABLE OF CONTENTS

INTRODUCTION

The investigations in this book were created to engage students in exploring the many mathematical ideas that can be discovered using the set of Power Polygons.

The investigations were designed to be used in many different types of classroom settings:

- A selection of student pages can be handed out one at a time or copied and bound to make a student booklet. Since students most often write their responses in notebooks or on separate sheets of paper, the copied pages can be saved to be used another year.

- The investigations can be used in an independent math center environment, for collaborative group work, or in a whole-class setting.

- Because the investigations within each chapter become progressively more difficult, individual students or groups of students can be challenged at an appropriate pace by allowing them to work independently.

- Students can respond to the numbered activities in each investigation by writing on separate sheets of paper, by writing in a notebook or journal, by making a group report chart, or by participating in a whole-class discussion.

- The investigations can be done in sequence over 30 to 40 days of instruction. Or, they can be done selectively throughout the year as they fit into your instructional plan.

When given the Power Polygons for the first time, most students will immediately start to produce symmetrical tiling patterns or representations of figures such as rockets, robots, or flowers. Since students will be exploring important ideas of congruence and shape recognition, such activities have mathematical as well as artistic merit. The investigations also focus on "free exploration," important mathematical ideas, including many that students are not likely to discover without some guidance.

Managing the Polygons

There are 450 pieces in the Power Polygons set, including 15 different shapes. The shapes are identified with the letters A through O. One set of Power Polygons is sufficient for a large group of students or a math center. Two sets will offer a whole class of students enough shapes for extensive explorations. The translucent quality and rich colors of the Power Polygons make them particularly exciting overhead projector materials. And, if students put one shape on top of another to compare angles, areas, or side lengths, they will be able to see both of the shapes at once.

Some teachers like to sort the polygons into 15 different containers labeled A through O. If this is done, it is easier to distribute specific shapes such as all the triangles. Students may often need one of each of the 15 shapes to begin an investigation. The needed shapes can be quickly distributed if each student makes a "Personal Power Polygons Set" by putting the 15 shapes in a zip lock bag and labeling the bag with his or her name. Another way for students to quickly select one of each of the 15 shapes is to use the Power Polygons Workmat and select shapes until each polygon is covered.

The polygons lettered A–F can each be covered by one or more F polygons and thus have many equivalence relationships. These polygons can be traced onto the square dot paper workmat. (One-inch graph paper can also be used.) The G–O polygons are different in that they are all related to the N polygon, the small equilateral triangle. These shapes can be traced onto the triangular dot paper workmat.

The back of the book has blackline masters for the three workmats: the Power Polygons, triangular dot paper, and square dot paper. Students can use the workmats as recording sheets to make investigating and reporting easier. Suggestions for using the workmats are offered in particular investigations, but you may wish to use them more often.

USING THE STUDENT INVESTIGATION PAGES

The four chapters are named for the content that is investigated: Logic and Patterns, Geometry, Perimeter and Area, and Fractions. The chapters can be completed in any order.

Each chapter has seven investigations which are best done in order. Each investigation is on a single page and is composed of several activities organized into three parts:

- Starting the Investigation begins the exploration of a concept with experiences such as sorting, combining, and reflecting.
- Going Further encourages students to consider additional aspects of the concept.
- Reflecting and Extending provides more opportunities for students who have time and interest in thinking more deeply about the ideas raised in earlier activities in the investigation.

It is usually best if each student has a copy of an investigation to read independently. Some students may need to focus on just one activity at a time. If so, they can fold the investigation horizontally so they have just the first activity to read. When the first activity is complete, students consider the next section of the page.

Students should record their findings on a separate sheet of paper or in a journal. Encourage them to describe specifically how they did the activities. A student response such as "I tried each piece" gives little information. To encourage more description, give students guidelines such as "I expect at least three sentences about your process."

Students who have difficulty writing can work with partners who are responsible for recording; however, both partners must do the thinking. Changing partners a few times over the course of all the investigations in the book seems to reduce the dependency factor that can arise in partnering.

Students will need different amounts of time for the investigating and the writing, so you may want to have some students find all possible solutions while others provide just a sample. This helps to keep everyone working on the same investigation and allows all students to participate in group discussions. If students are working through the investigations independently, it is important to keep them from rushing ahead without careful, deep thinking. Set high standards for the full completion of each investigation when students are working on their own.

USING THE TEACHER PAGES

For each student investigation page there is one teacher page. The teacher pages offer many tips for responding to students' work. They also include management suggestions and information about what materials are needed for each investigation.

Solutions to the activities are given on the teacher pages. When an activity is open in nature and may have alternative answers, the mathematics of the activity is explained so that you can make informed decisions about the reasonableness of students' solutions.

Be sure to read the teacher page before assigning the student investigation. You may also wish to try the investigations yourself. Students sometimes think of possibilities that have not been discussed on the teacher page. Having done an investigation yourself may give you insights you can use to respond.

ONE CLASSROOM EXPERIENCE

Many of these investigations were tried during the first month of school in a typical fourth grade classroom. Each child found one of each of the 15 different polygons and then kept them in a zip lock bag for the duration of the investigations. The remaining polygons were available in the math center for use during free time. The 15 polygons per child were sufficient for most of the investigations; larger quantities were rarely needed.

Students were organized in groups of four and encouraged to ask each other for explanations and ideas before calling on the teacher. Generally, students were able to complete about one investigation in an hour's math period. Students who needed less time were allowed to work independently with the Power Polygons to design their own patterns and investigations.

Little introduction was given to each investigation so that students were encouraged to explore the activities in their own way. However, expectations were carefully outlined. These included such things as helping someone when asked, writing in complete sentences, keeping all papers until the final draft was copied in the journal, drawing polygons with rulers so the edges were straight, and sharing experiences with parents. There was a discussion session after each investigation to capture the richness of the different perspectives that individuals brought to the tasks.

Students were expected to write a draft report for each investigation. One day a week was reserved for revising work before the final reports were copied into students' journals. Students based their revisions on teacher feedback or comments made by other students during class discussion. Students were always encouraged to improve and revise their work by building on the thinking of others.

Though many students in the class had attention problems, they were very involved in the investigations and continually improved in their ability to work without teacher support. Students also improved in recording their thinking in writing. In almost all cases, the students amazed us with their interest and insight. They truly enjoyed using the materials and took pride in the products of their efforts.

Logic
and
Patterns

THE SET OF POWER POLYGONS

GOAL OF THE INVESTIGATION

Students analyze the attributes of the 15 different pieces that form the total set of Power Polygons.

FACILITATING THE INVESTIGATION

If students are having difficulty, suggest that they place the pieces on top of one another and then turn the pieces to determine if they are the same or different.

   If time allows, encourage students to make designs with the polygons, since this seems to be something they are drawn to do. Not only will this satisfy their need to explore the creative possibilities, but it will help them develop an intuitive awareness of relationships among the shapes.

SAMPLE SOLUTIONS AND DISCUSSION

 There are 15 different polygons in the Power Polygons set. Some students may indicate that they know they have found all the pieces by saying "I just checked all the pieces." You might ask them what they think would happen if they sorted the entire set into piles with each pile having the same pieces. If they do this, they may begin to discuss the attributes of the shapes. They may also notice that each different shape has a unique letter, A–O.

 Most likely students will notice that the set includes polygons with 3 sides, 4 sides, and 6 sides. They may know the names of some of the shapes, but not necessarily all of the shapes. Including the names *triangles*, *quadrilaterals*, and *hexagons* in the discussion may make it easier to talk about the pieces.

REFLECTING AND EXTENDING

Throughout the investigations in this book, students will be asked to describe their thinking about geometric ideas that emerge as they work with the polygons. This first investigation is a good place to set standards for responding to open-ended questions.

 Responses to Activity 2 in this investigation will probably be an informal list. In Activity 3, students are more likely to write in sentences. Accurate representation of mathematical ideas with pictures is helpful. Since polygons must have straight sides by definition, encourage students to use a ruler or to trace the shapes on dot paper so that the sides of the polygons are made from straight line segments.

The Set of Power Polygons

Starting the Investigation

Take a good look at the Power Polygon pieces.

a. How many different shapes can you find in the set?

b. How do you know you have found one of each different shape?

Going Further

a. What do you notice about this set of shapes?

Make a list of what you might say in describing the polygons.

b. Work with a partner. Discuss your lists and work together to make a final list.

Reflecting and Extending

Work with your partner. Use your final list to create a poster that describes the Power Polygons. You can use pictures and words.

2 LIKENESSES AND DIFFERENCES

GOAL OF THE INVESTIGATION

Students create sorting rules and then sort the pieces according to their rule. This gives students the opportunity to closely analyze the similarities and differences among the shapes.

FACILITATING THE INVESTIGATION

Before students start the activities, hold up two items such as two different books or two different sneakers. Ask students to think of terms commonly used to describe distinguishing features. They should suggest such attributes as: shape, color, weight, thickness, etc. Record these terms on the chalkboard so students can refer to them as they do the activities.

For Activity 2, consider having students compare more than one pair of polygons. This may help broaden their list of characteristics. Similarly, having them sort the set of polygons in more than one way encourages students to look for many different solutions, even when they are asked for only one.

SAMPLE SOLUTIONS AND DISCUSSION

 Many answers are acceptable here. The most common attributes that students use to describe likenesses and differences are color, the number of sides, and the type of polygon.

Discussion of possible likenesses and differences could also touch on a comparison of the area, symmetry, and size of the angles. For example, students might note that one polygon is a fractional part of another, or they may discuss that one shape has small corners that are "pointy" while the other has "square corners." These concepts are investigated in later activities.

A way to extend and compare diverse responses is to have students trade papers and ask their partners to add to the likenesses and differences of the polygons that were chosen.

 Students should be inclusive in their sorting, that is, they should not leave pieces out that meet their sorting criteria. Having partners reflect on each other's sorts helps children to verbalize their sorting rules.

Students at this age should be able to sort the set of polygons several different ways. Challenge them to create sorts that require the use of two attributes in describing each group, for example: 4-sided blue shapes or large equal-sided shapes.

REFLECTING AND EXTENDING

 This extension can help validate that students understand the attributes they are using, since they will be creating, and not just selecting, polygons. Students may want to color the shapes they make and add them to their groups of polygons.

 Likenesses and Differences

Starting the Investigation

Choose two different polygons and trace them.

 a. How are the two shapes the same?

b. How are the shapes different?

Going Further

Sort a handful of the polygons into two different groups.

 a. What is different about the two groups? Write about what you did.

b. Work with a partner. Discuss the ways each of you sorted your pieces.

Reflecting and Extending

 Use a ruler and dot paper to draw new polygons that could go into your two groups.

VENN DIAGRAMS

GOAL OF THE INVESTIGATION

Students explore Venn diagrams as one type of graphic organizer that can be used to show ideas pictorially. Students should recognize that not all sorts are discrete (that is, have nothing in common).

FACILITATING THE INVESTIGATION

It is beneficial for students to learn different organizational techniques that help in finding patterns and solutions. The Venn diagram is a clever way to resolve representing overlapping information; however, students may have other solutions that also solve the problem. Be sure to encourage their independent investigation of this dilemma before helping students with Venn's approach.

You may need to discuss the meaning of overlapping circles. If students have difficulty grasping the technique of sorting into two overlapping circles, be sure they label each circle and then place the actual shapes in the circles. They should then be able to record the letter names when the sorting has been completed. It will be simpler for students if they work from a single set of the 15 unique polygons rather than a larger set.

SAMPLE SOLUTIONS AND DISCUSSION

 Acceptable solutions include placing the N and E pieces on the fold line of the paper or creating a third space for those polygons that have both attributes. Students should label the sets so that their thinking can be understood by anyone looking at their work.

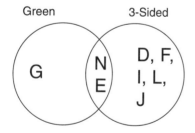

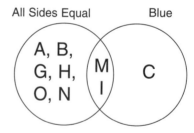

REFLECTING AND EXTENDING

 Creating their own Venn diagram can be more difficult for students than representing one that is posed to them. Having partners try each other's puzzles is a good way to verify that the logic used in creating a puzzle makes sense.

Venn Diagrams

Starting the Investigation

Fold a piece of paper into two parts. Put green polygons on one side and 3-sided polygons on the other.

You may not repeat a letter in this sorting.

 What did you do with the N piece? the E piece?

Going Further

A mathematician named Venn used two overlapping circles for sorting shapes when some shapes belonged in both groups.

 What might Venn's diagram for Activity 1 look like?

Draw a diagram. Use the letter names for the polygons.

 How might Venn have shown these two groups?

Polygons with all sides equal   Polygons that are blue

Reflecting and Extending

 Create a Venn diagram puzzle for your classmates by drawing a picture of a sorting of some of the polygons.

Don't include labels this time—that will be your classmates' challenge!

MORE VENN DIAGRAMS

GOAL OF THE INVESTIGATION

Students continue to explore the use of Venn diagrams in conjunction with the word "not." This activity gives students the opportunity to organize information in Venn diagrams at a more challenging level.

FACILITATING THE INVESTIGATION

If students are unsure about how to begin, suggest they start by removing all 4-sided shapes from the full set of 15 Power Polygons. Remind them, however, to check this removed set when they look for blue polygons.

You may also need to direct students who aren't able to organize to draw two large overlapping circles and to label them.

SAMPLE SOLUTIONS AND DISCUSSION

 Ask students to describe why they placed each set of polygons in a particular section of the diagram. The polygons in the left circle are not quadrilaterals. The polygons in the right circle are blue. The polygon in the intersection of the circles is not a quadrilateral and it is blue.

NOTE: The answer to the puzzle is M, O, and G.

 The polygons in the left are green and the polygons in the right circle are polygons that have all sides equal. The set in the intersection is both green and equilateral.

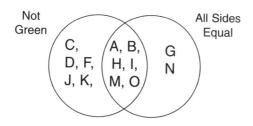
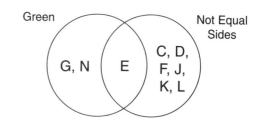

REFLECTING AND EXTENDING

 Students should see that there is no overlap, *intersection*, when a set is shown with its complement. For example, these two groups are complements: the set of triangles and the set of polygons that are not triangles. The reason these sets are complements is that, together, they describe all the polygons.

 Students may begin to notice that sometimes use of the word "not" results in two groups with no intersection because there is no polygon that has both attributes. At other times the result does include an intersection because there is at least one polygon that belongs in both categories.

More Venn Diagrams

Starting the Investigation

Use only one of each type of polygon.

 What would the sort of these two groups look like?

Not quadrilaterals
(not 4-sided shapes)   Blue polygons

Going Further

 What label should go above
each of these two circles?

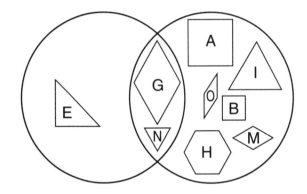

 Show how the sorting changes if the word *not* is put in
front of one of the labels.

Reflecting and Extending

 If you used a Venn diagram to show *triangles* and *not
triangles* what would it look like?

 How does including the word *not* change the sorting of polygons?

POLYGON PUZZLES

GOAL OF THE INVESTIGATION

Students explore the attributes of different polygons by solving and creating puzzles. They become familiar with the problem solving technique of eliminating impossible answers.

FACILITATING THE INVESTIGATION

Demonstrate the game first. Cover your eyes as the class watches one student hide a related set of polygons. Model good problem solving by asking "yes/no" questions that reduce the possibilities as quickly as possible.

FOR EXAMPLE: Do the polygons have more than 3 angles?
Are more than 2 sides equal in length?

Then reverse roles by having the students ask questions about a set of polygons that you have hidden.

Have the students brainstorm some of the sets of polygons that they could hide. Recording these ideas on the chalkboard will help those students who have difficulty thinking of a set to hide. It could also be helpful to students who are trying to find a possible solution.

Large index cards are good for making the puzzle cards for Activities 2 and 3.

SAMPLE SOLUTIONS AND DISCUSSION

 Students having difficulty may benefit by hearing successful students share their techniques and by seeing their written responses.

 Since there are so many possibilities for Activity 2, this is a place where peer editing can be helpful. Tell the students they must have at least 2 students successfully solve their puzzle before it can go into the class collection.

Puzzles can be reproduced for a homework sheet. Students can take home a copy of the Power Polygons Workmat (page 64) to help find the solutions. An alternative is to put the puzzles in a activity center for students to solve during unassigned time.

REFLECTING AND EXTENDING

 Students extend their skills involving "asking good questions." They apply what they have learned using the polygons to the objects in the world around them.

Answer to puzzle {M, O, G}

Polygon Puzzles

Starting the Investigation

Have your partner hide a set of polygons that has one or two special attributes.

Guess the attribute(s) of your partner's polygons by asking *yes* or *no* questions about the shapes.

a. What questions helped you identify the attributes?

b. What polygons are in your partner's hidden set?

Going Further

Try this puzzle. Remember, the group of polygons must match *ALL* the clues.

> ### CLUES
>
> The polygons have no right angles (angles that match the corners of a square).
>
> They have four equal sides.
>
> What polygons are they?

a. Select a set of polygons. Use clues, like those above, to create polygon puzzles for the other students.

b. For the puzzles you create, put the clues on one side of a piece of paper and the answers on the back.

c. Give your puzzles to another student to read and edit if necessary.

Reflecting and Extending

You can use the same techniques with objects in your classroom.

a. Ask your partner to hide something and respond to your *yes* or *no* questions about the attributes of the hidden object.

b. Write down the clues you discover so you can turn them into a puzzle for other students to try.

6 POLYGON TABLES

GOAL OF THE INVESTIGATION

Students explore and continue patterns made by adjoining polygons. They may discover a generalization that will help them find the solution given any number of polygons.

FACILITATING THE INVESTIGATION

Some students will need to build all the arrangements to find the solution for 10 polygons. Others may be able to discover a generalization or even abstract a rule. Encourage students to work with partners or groups. This cooperative effort can broaden students' perspectives as well as assure there are enough polygons for modeling.

SAMPLE SOLUTIONS AND DISCUSSION

 Students who collect their data in a chart are likely to notice that the adjoining of a new triangle adds one additional seat each time.

tables	1	2	3	4 . . .	10
students	3	4	5	6 . . .	12

This generalization works well when you are working with small quantities.

When trying to determine how many seats will be created by 100 triangular tables, some students will try to record all the numbers through 100. Other students may make the generalization "the number of tables plus 2." Written algebraically this generalization is "n + 2." However, the word phrase may be more meaningful to students at this age.

If children discover related ideas, allow them to investigate them at another time. For example, if 6 triangles are joined to make a hexagonal table rather than adjoining one side, the rule above does not work because more of the sides are enclosed leaving 6 seats rather than 8.

 This time, each additional table adds 2 seats and the generalization is "the number of tables times 2 plus 2" or "2n + 2."

tables	1	2	3	4 . . .	10
students	4	6	8	10 . . .	22

Encourage students who find the rule quickly to keep it a secret until everyone has had the chance to find a solution.

REFLECTING AND EXTENDING

 All quadrilaterals work like the square if one seat is on each side.

Polygon Tables

Starting the Investigation

Pretend the cafeteria tables are shaped like
polygon N and one student can sit on each side.

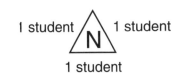

1 student · 1 student
N
1 student

 a. Suppose you pushed these triangular tables together so
one side touches another. Then how many students
could sit at 1 table? at 2 tables? at 3 tables? at 4 tables?
at 10 tables?

  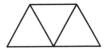

Record your findings.

b. What patterns do you notice?

c. How many students could sit at 100 tables?

Going Further

 a. Explain what happens if the tables are shaped like
polygon B.

b. How many students could sit at 10 square tables that
touch on one side?

Reflecting and Extending

 Investigate what will happen if you join any of the
4-sided polygons.

SQUARE PATTERNS

GOAL OF THE INVESTIGATION

Students discover a rule for the special set of numbers known as square numbers. They also explore other patterns generated by adjoining squares.

FACILITATING THE INVESTIGATION

Give students copies of graph paper or square dot paper (page 66) for these activities. This will allow them to keep a record of their findings.

Students working in pairs or groups are more likely to be successful in eliminating duplicate answers in Activity 2.

SAMPLE SOLUTIONS AND DISCUSSION

  If students don't see the solution to a 10-sided square, have them continue to fill in the chart by shading squares on graph paper for 6–9 squares on a side.

Total number of squares	1	4	9	16	25 . . .	100
Number of squares on a side	1	2	3	4	5 . . .	10

Most students who have studied multiplication facts will eventually observe that the total number of squares can be obtained by multiplying the length of one side of a square by the other.

Encourage students who do not see the connection to consider the number of rows, and the number in each row in their models. The pattern of two rows of two, three rows of three, . . . , ten rows of ten, should become apparent to them. To find the total number of squares they will probably suggest something similar to: "add the number of squares in a row that many times" or "count by the number in each row."

  This collection of arrangements is called *tetrominoes*. Encourage students to eliminate duplicates, those arrangements that are flips or turns of each other.

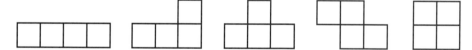

REFLECTING AND EXTENDING

  These are the *pentominoes*. There are 12 altogether. Generally, students use trial and error to investigate these. One approach to finding them is to start with each of the tetrominoes as a base and to consider the new arrangements that are possible by adding the additional square. You can identify the pentominoes by the letter names they appear most like. F, N, and Y need more imagination than the others.

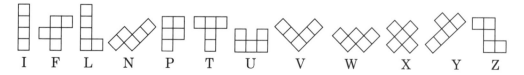

Square Patterns

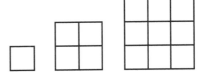

Starting the Investigation

Make larger squares by connecting smaller squares.

a. How many of the small squares do you need to make a larger square with 2, 3, 4, and 5 squares on a side?

b. How many small squares would you need to make a large square arrangement with 10 small squares on a side?

Going Further

Take 4 squares, but this time, put them together so that at least one side of each matches up exactly with the side of another.

a. What different arrangements can you make in this way with 4 squares?

Check for any that are the same. Try to picture how the arrangment will look if it is turned or flipped. Make a cutout to help you.

b. What strategy did you use to find different arrangements?

Reflecting and Extending

Find all the different arrangements that can be made with 5 squares. Show your answers on graph paper or square dot paper.

(Hint: There are more than 10.)

CHAPTER 2

Geometry

◣1 SIDES AND ANGLES

GOAL OF THE INVESTIGATION

Students explore the attributes that are typically used to distinguish different types of polygons.

FACILITATING THE INVESTIGATION

Provide students with square dot paper (page 66) for drawing polygons A–F and triangular dot paper (page 65) for polygons G–O.

If students are consistently creating new polygons with 3 or 4 sides, ask them if they could draw polygons with a different number of sides, such as 5, 6, 8, and so forth.

SAMPLE SOLUTIONS AND DISCUSSION

◣1-2 Power Polygons A through O can be sorted into three general types of polygons.

triangles	quadrilaterals	hexagon
D E F I	A B C G	H
J L N	M O K	

Other attributes that students notice may include:
- some of the polygons are regular, with all angles equal
- others have a pair of equal angles, or no angles equal
- a side on each polygon matches with a side on at least one other polygon
- the lengths of the sides measure 1, 2, or 4 inches, with the exception of some sides of triangles
- the number of angles and the number of sides are the same

The most important attribute for students to understand is that polygons are made from connected, straight line segments.

Have students identify any of the names of the polygons that they know. Triangle, quadrilateral, pentagon, hexagon, heptagon, octagon, nonagon, decagon, undecagon, and dodecagon are the most common.

REFLECTING AND EXTENDING

 For a polygon to be made from connected straight line segments, there must be at least three sides. Triangles are the polygons with the fewest sides.

A new polygon made by adjoining 2 hexagons is a decagon. Other decagons are possible, but 10 sides are the maximum with two of the Power Polygons.

Sides and Angles

Starting the Investigation

Take one of each polygon A through O.
Sort the polygons by the number of corners.

a. What do you notice about the corners, which are also called angles?

b. What else do you notice about the polygons in each of your sets?

Going Further

Make new polygons by putting two polygons together and tracing the outside of the new shape on dot paper.

a. What can you say about the angles of the new polygons?

b. What else do you notice about the new polygons you've made?

Reflecting and Extending

a. What is the fewest number of sides a polygon can have?

b. What combination of two polygons has the most number of sides? Draw some examples of these polygons.

MORE SIDES

GOAL OF THE INVESTIGATION

Students explore creating different polygons that are concave and convex.

FACILITATING THE INVESTIGATION

Provide students with both kinds of dot paper (pages 65 and 66).

Sometimes students have only seen polygons represented by shapes that are convex, no indents, and that have all equal sides. Displaying different examples of polygons on large sheets of paper can help students appreciate the many possibilities that exist. An alternative is to have each group produce a Book of Polygons with a page dedicated to each type.

SAMPLE SOLUTIONS AND DISCUSSION

 Some possibilities for concave polygons are:

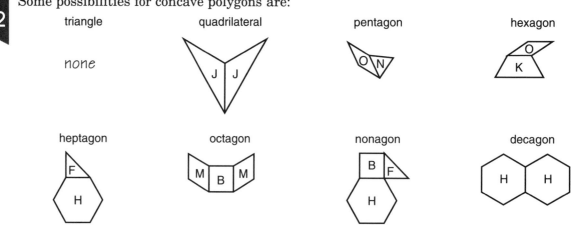

 A concave triangle is not possible. A concave figure has at least one angle greater than 180°; but, the sum of all the angles in a triangle is only 180°. Students may say you need more than 3 straight sides or at least 1 curved side to close a concave figure.

REFLECTING AND EXTENDING

 As students work with dot paper they may create illustrations that are not polygons. There should be no intersections and no holes in the interior of a polygon.

Answers will differ based on the dot paper used. One student was able to create a 66-sided polygon. Keeping a record of the most number of sides gives you the opportunity to provide the children with a benchmark. "Leah created a concave polygon with 20 sides. Is it possible to make a concave polygon with more sides?"

EXPLORING WITH POWER POLYGONS
©1996 Cuisenaire Company of America

More Sides

Starting the Investigation

Join polygons to make a new set of polygons.

 Trace them onto dot paper. Include a polygon with 3 sides, 4 sides, 5 sides, 6 sides, 7 sides, 8 sides, 9 sides, and 10 sides.

Going Further

Some of the polygons you created may have indents.

Polygons with indents are called *concave*.

Polygons that do not have indents are called *convex*.

concave

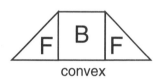

convex

 a. Write concave under each polygon that you made that has indents.

      b. Write convex under each polygon that you made that does not have indents.

 Can you create a concave triangle? Why or why not?

Reflecting and Extending

 Show a concave polygon with the most number of sides. Use the polygons or dot paper.

ANGLES

GOAL OF THE INVESTIGATION

Students explore the size of angles by direct comparison to a right angle.

FACILITATING THE INVESTIGATION

Students should begin with one of each of the 15 unique polygons. Students can use the corner of an index card or a piece of paper as a right angle for comparison rather than the square polygon B. This eliminates the confusion about which polygon is the measuring device, and which is being measured. Students having difficulty may benefit from using the Power Polygons Workmat (page 64).

SAMPLE SOLUTIONS AND DISCUSSION

 A, B, C, D, E, F, and L have an angle that matches an angle of a square.

 G, H, J, K, M, and O have an angle that is bigger than an angle of a square.

I and N have angles that are all smaller than the angles of a square.

Some students may be familiar with the commonly used names of the different-sized angles.

> Angles of a square are called *right* angles.

> Angles smaller than a square's angles are called *acute*.

> Angles greater than a square's angles are called *obtuse*.

REFLECTING AND EXTENDING

 This investigation is a precursor to understanding the degrees that are used to measure the size of angles. Students will not know the degrees but can find solutions by experimentation.

Some solutions for surrounding a point are:

J, L, O	12 of the 30° angles
D, E, F	8 of the 45° angles
G, I, K, L, M, N	6 of the 60° angles
A, B, C, D, E, F, L	4 of the 90° angles
G, H, J, K, M	3 of the 120° angles

NOTE: Polygon O has a 150° angle which will not evenly fill the space around a point.
    Also, combinations of polygons will fill the space around a point, for example: 2 J and 2 M polygons will work, since 2 x 120° + 2 x 60° = 360°.

Angles

Starting the Investigation

One way to sort shapes is by the size of their angles.

 Look at each polygon A through O.

Which polygons have at least one angle that matches the angles on the square polygon B?

Going Further

 a. Which polygons have an angle bigger than those on polygon B?

b. Which polygons do not belong in either group and what can you say about their angles?

Reflecting and Extending

It takes 3 of the large angles on the M polygons to fill the space around a point.

 a. What happens if you surround a point with the small angles on polygon M?

b. How many of each of the other polygons are needed to fill the space around a point?

SYMMETRY

GOAL OF THE INVESTIGATION

Students explore lines of symmetry using the Power Polygons.

FACILITATING THE INVESTIGATION

Encourage students who are confused about the symmetry of a polygon to test it by cutting out a tracing and folding it. The Power Polygons Workmat (page 64) can be used instead of tracings.

SAMPLE SOLUTIONS AND DISCUSSION

 Only L has no line of symmetry.

One line of symmetry	D, E, F, J, K
Two lines of symmetry	C, G, M, O
Three lines of symmetry	I, N
Four lines of symmetry	A, B
Six lines of symmetry	H

REFLECTING AND EXTENDING

 The hexagon has the greatest number of lines of symmetry.
    After seeing regular polygons with increasing numbers of sides, some students may be able to generalize that as the number of equal sides in a shape increases, the number of lines of symmetry increases.

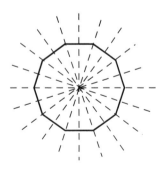

    If we consider a polygon with a hundred equal sides fitting on a piece of standard paper, it would appear almost circular and have 100 lines of symmetry. A circle, in fact, has an infinite number of lines of symmetry.
    Figures with no equal sides are likely to have no lines of symmetry.

Symmetry

Starting the Investigation

Polygons have symmetry if they fold along
a line to make two matching parts.

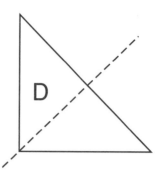

 Which polygons have lines of symmetry?

Going Further

 a. Which polygons do you think have more than one line
of symmetry?

b. Show these polygons with their lines of symmetry.

Reflecting and Extending

 a. Which polygon has the most number of lines
of symmetry?

b. Create polygons that have even more lines of symmetry.

c. Create polygons that have no lines of symmetry.

5 SYMMETRICAL PUZZLES

GOAL OF THE INVESTIGATION

Students explore symmetry using a combination of shapes. They create symmetrical designs.

FACILITATING THE INVESTIGATION

Each student will need 3–4 of each type of polygon.

You may wish to restrict the symmetrical designs to the size of a single piece of dot paper.

To make the task simpler, a student can fold a piece of dot paper in half and then open it up. Then the student creates a design against the fold and trace it. Finally, the student copies or creates the other half of the design. A mirror can help a student reflect the design over the line of symmetry.

To make the task more challenging, ask students to create designs with more than one line of symmetry.

SAMPLE SOLUTIONS AND DISCUSSION

 Students should be able to illustrate the line(s) of symmetry in their design.

 Students must be able to reproduce their own designs from their outlines so that they can help other students who will be trying them.

If you want an answer key for each puzzle, have each student draw in all the shapes on a copy of the outline of his or her design.

Covering a wall or a bulletin board with colored copies of all the designs produces a beautiful quilt effect.

REFLECTING AND EXTENDING

 Helping students see how the mathematics they are learning applies to the real world makes it more meaningful. This activity encourages students to find such connections. Architecture, decorative arts, and wall coverings are rich with examples of symmetrical designs. Leaves and flowers also provide samples.

The search for symmetrical designs can be made more expansive by assigning teams of students to particular areas of the school for a "Symmetry Hunt." Have them share their findings with the whole class.

EXPLORING WITH POWER POLYGONS
©1996 Cuisenaire Company of America

Symmetrical Puzzles

Starting the Investigation

A symmetrical design can be made with a combination of shapes.

a. Make a symmetrical design from several of the polygons.

b. Trace the perimeter to draw an outline of your design.

c. List the type and number of each polygon that you used in your design.

Going Further

a. Remove the polygons and make sure you can recreate the design on top of the drawing.

b. Trade puzzles with a partner to see if you can do each other's puzzles.

Reflecting and Extending

Symmetry is everywhere around us both in natural and machine-made objects.

a. Write a description or draw a sketch of some symmetrical items that you see around you.

b. Use the Power Polygons to create a picture of an item that is symmetrical and made from polygons such as a necktie.

6 TRIANGLES

GOAL OF THE INVESTIGATION

Students explore the attributes of triangles by sorting them in their own way. Then, they sort them by equal side lengths.

FACILITATING THE INVESTIGATION

Suggest that students isolate one of each type of triangle from the other polygons. They can place the triangles on top of one another to make sure that they have not included any duplicates.

Some students may have difficulty determining attributes by which they can sort the triangles. Encourage them to identify some of the triangles that have something in common and put them into one group. Have the students record the common attribute of this group. The remaining triangles can be described as those triangles that do not have this attribute.

SAMPLE SOLUTIONS AND DISCUSSION

 There are many ways to sort the triangles. Color, angle or side measure, and triangles that are one half of another polygon are some of the possible attributes that may be used. You may wish to have each group show unique possibilities on an overhead projector.

 Equilateral (three equal sides): N, I
Isosceles (two equal sides):   D, E, F, J
Scalene (no equal sides):   L

 Two isosceles triangles can be made by halving: A, B, D, E, F, G, M, O
Two scalene triangles can be made by halving: C, I, J, N

REFLECTING AND EXTENDING

A square, a rhombus, and a right isosceles triangle fold in half to make isosceles triangles. Students may discover that if they fold down the long side of a sheet of paper so that one side touches the other they have made a right isosceles triangle. This is one half of a square. If they tear off the extra at the end of the sheet, they will have made a square.

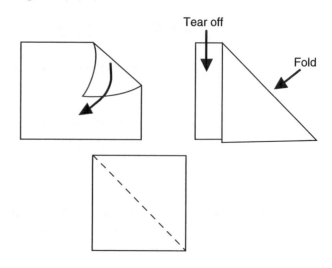

 Triangles

Starting the Investigation

Use these six pieces: D, E, F, I, J, L, and N.

 a. Sort the triangles into different groups.

b. How would you describe each group of triangles?

 a. What do you notice about the side lengths of the different triangles?

b. Sort the triangles by the number of equal sides.

Going Further

Use all the different types of polygons in the Power Polygons set.

 a. Which polygons can be cut in half to make triangles that have 2 equal sides (isosceles)?

b. Which polygons can be cut in half to make triangles with no equal sides (scalene)?

Reflecting and Extending

 a. What polygons can be folded in half to make isosceles triangles?

b. Use this information to find a way to fold a sheet of paper into a square.

QUADRILATERALS

GOAL OF THE INVESTIGATION

Students explore the attributes of quadrilaterals and create different types of quadrilaterals based on the number of parallel sides.

FACILITATING THE INVESTIGATION

Suggest that students isolate one of each type of quadrilateral from the other polygons. They can place the quadrilaterals on top of one another to make sure that they have not included any duplicates.

Some students may have difficulty determining attributes by which they can sort the quadrilaterals. Encourage them to identify some of the quadrilaterals that have something in common and put them into one group. Have the students record the common attribute of this group. The remaining quadrilaterals can be described as those quadrilaterals that do not have this attribute.

SAMPLE SOLUTIONS AND DISCUSSION

 There are many ways to sort the quadrilaterals. Some include: color, number of equal sides or angles, those with right angles, or those that look like diamond shapes. You may wish to have each group show unique possibilities on an overhead projector.

Some students may notice that the number of parallel sides and the number of sides with equal lengths distinguishes one quadrilateral from another. Their names are squares, rectangles, rhombuses (diamond shapes), parallelograms, and trapezoids. At this grade level, students need only appreciate that parallelograms are special trapezoids with two pairs of parallel sides. Rectangles, rhombuses, and squares are all special types of parallelograms.

 Polygon K (the trapezoid) has only one pair of parallel sides. Polygons A, B, C, G, M, and O all have two pairs of parallel sides.

 Some possible solutions:

One pair of parallel sides: AE, BF, BL, CL, EC, IL

Two pairs of parallel sides: LL, OO, AA, DD, AC, BC

REFLECTING AND EXTENDING

 Some polygons can be combined to make new quadrilaterals that have no parallel sides. Some possibilities are: NF, EL, EJ, FN, JJ.

 The rectangle seems to be the most visible of the polygons probably because of its ability to join to and support other shapes.

Quadrilaterals

Starting the Investigation

Use these seven pieces: A, B, C, G, K, M, and O.

a. Sort the polygons with 4 sides (quadrilaterals) into different groups.

b. What do you notice about the different types of quadrilaterals?

Going Further

a. Which quadrilaterals have one pair of parallel sides?

b. Which quadrilaterals have two pairs of parallel sides?

Use all the different types of polygons in the Power Polygons set.

a. Which polygons can be put together to make quadrilaterals with only 1 pair of parallel sides?

b. Which polygons will make quadrilaterals with two pairs of parallel sides?

Reflecting and Extending

Use the Power Polygons to create some quadrilaterals with no parallel sides.

Which type of quadrilateral do you see most often in the world around you? Why do think that is so?

Perimeter
and
Area

DOUBLES AND QUADRUPLES

GOAL OF THE INVESTIGATION

Students cover the surface of one Power Polygon with two or more of another. They are exploring the idea of area as covering a surface.

FACILIATING THE INVESTIGATION

Students who have difficulty covering one shape with others may find using the Power Polygons Workmat (page 64) simplifies the process. They can record their findings with sentences next to the shape.

Another way to simplify the search is to separate the set of 15 polygons into subsets. The A–E polygons have many equal area relationships because two or more small right triangles (F) will cover each of them. The G–M polygons can all be covered by two or more equilateral triangles (N). The rhombus (O) will not cover any of the other polygons.

SAMPLE SOLUTIONS AND DISCUSSION

 The doubles that can be joined to equally cover another Power Polygon are:

A = 2C or 2E	G = 2I or 2J
B = 2F	H = 2K
C = 2B	I = 2L
D = 2E	J = 2L
	M = 2N

Other doubles relationships require the use of logic or breaking polygons for their validation. They include:

E = 2B because A = 2E = 4B

I = 2M because G = 2I = 4M

J = 2M because G = 2J = 4M

L = 2N because G = 4L = 8N

 The apparent quadruples are:

A = 4B   G = 4M or 4L   C = 4F   I = 4N   E = 4F

Less apparent is the fact that J = 4N because G = 2J = 8N.

REFLECTING AND EXTENDING

 From the above mentioned relationships, it can be argued that the following statements can be made about areas of these polygons:

A = D   C = E   I = J   L = M

Statements that are easily proven are:

D = 8F   G = 8N   H = 3M = 6N   K = 3N

# Doubles and Quadruples

Starting the Investigation

Shapes are equal in area when they can be made to cover each other exactly. Some of the Power Polygons can be covered by 2 of another polygon. Call them *doubles*.

 What different doubles relationships can you find in the set of Power Polygons? Record with sentences or drawings.

EXAMPLE: 2 B = C

Going Further

Some of the polygons can be covered by 4 of the same shape. Call them *quadruples*.

 What different quadruple relationships can you find in the set of Power Polygons?

Reflecting and Extending

There are other equal area relationships among the Power Polygons.

 a. How many can you find that use multiples of one polygon to exactly cover another?

b. What sentences could be used to show these relationships?

MIXED COVERINGS

GOAL OF THE INVESTIGATION

Students continue to explore covering one shape with others. They are encouraged to find and explain ways to be complete in their search for solutions.

FACILITATING THE INVESTIGATION

A group working together might want to leave all the possibilities visible as they search for more. An alternative is to have students trace and color their findings on a sheet of triangular dot paper which can be found on page 65.

SAMPLE SOLUTIONS AND DISCUSSION

 One way to find all the possible apparent coverings for the hexagon is to begin with the smallest polygon and then replace two or three pieces at a time:

H = 6N	H = 4N + M	H = 3N + K	H = 2K
H = 2N + 2M	H = N + K + M	H = 3M	

Some students may want to include N + M + N + M because it is visually different from M + M + N + N (2M + 2N). They may also include the H itself because it does technically cover an H. So, there are 7, 8, or 9 different ways to cover the hexagon.

Students may argue that they have found all possibilities because only the K, M and N polygons have equal covering relationships with the hexagon. Be sure to press them for further justification, especially if they have not found all the possibilities. You may need to suggest that they develop a way of sorting their findings, for example, all the coverings with an N and all without. This may help them see the solution in parts rather than a large collection.

 One way to find all the coverings for rhombus G is to work from the largest polygons to the smallest.

2J = G	2I = G	4L = G	4M = G	8N = G

REFLECTING AND EXTENDING

 Polygons F, L, N, and O cannot be covered exactly by the others, generally, because they all have small areas. However, it can be argued that if the shapes could be cut up, that 2Ns would cover an L, since I = 2L = 4N.

Covering relationships can help make interesting designs. They make it possible to allow the substitution of several polygons for one when there are not enough of one kind of shape. They help describe and identify fractional relationships. Students are likely to have other interesting ideas about the value of working with ideas of area.

Mixed Coverings

Starting the Investigation

The Power Polygons can be covered exactly by using more than one type of polygon.

a. What are all the ways hexagon H can be covered exactly by other polygons?

b. How can you prove that you have found them all?

Going Further

What are all the ways you can cover the largest rhombus G exactly?

Reflecting and Extending

a. Which polygons cannot be covered exactly by any of the other polygons? Why?

b. Why might it be helpful to know about equal covering relationships?

3 UNITS OF AREA

GOAL OF THE INVESTIGATION

Students use a small triangle as a unit for area measure. As they create new polygons with a given area, they are exploring the idea that many different shapes can have the same measure. They also explore ideas of equality by rotation (turn) and reflection (flip) as they design unique polygons of a given area.

FACILITATING THE INVESTIGATION

If students are having difficulty visualizing the area of the polygons in triangular units, they may find it helpful to trace them on triangular dot paper (page 65).

Some students may quickly discover that only the G–N polygons can be covered exactly by a whole number amount of Ns.

SAMPLE SOLUTIONS AND DISCUSSION

1. By direct measure with N (small green triangle), the areas of the other polygons are:

   G = 8 units   H = 6 units   I = 4 units   K = 3 units   M = 2 units   N = 1 unit

2. Indirectly students may discover that J = 4 units because it is half of G and L is 2 units because it is half of I.

3. The following combinations will yield totals of area 10:

   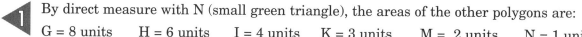

   | H + I | H + 2M | H + K + N | H + M + 2N | H + 4N | 2K + I |
   | 2K + 2M | 3K + N | 2K + M + 2N | 2K + 4N | G + M | G + 2N |

   Some of these combinations can take more than one shape, depending on how they are arranged.

4. The important idea here is that shapes that appear different visually can be equal in area because they can be covered by the same number of units.

REFLECTING AND EXTENDING

5. If they are having difficulty determining whether polygons are different, have students cut out the polygons they create. They will then be able to turn and flip the shapes when looking for duplicates.

   There are only four unique arrangements of 5N units when vertices and sides touch.

   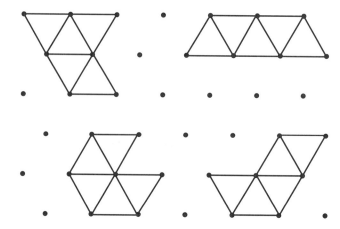

Units of Area

Starting the Investigation

Figuring out how many of one shape covers another is called finding the area of the larger shape. The smaller shape is the unit used to measure that area.

EXAMPLE: When triangle N is the unit of measure, hexagon H is 6 units in area.

 What are the areas of the other Power Polygons that can be covered exactly by collections of Ns?

 How could you determine the areas of J and L, in N units, without actually covering them with Ns?

Going Further

 a. What new polygons with an area of 10N units can you make by joining Power Polygons together on a side so that the corners touch?

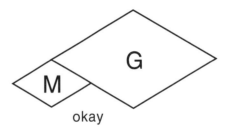

okay

b. Record polygons on triangular dot paper with corners at the dots.

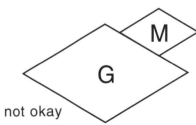

not okay

 How are the polygons you made alike?

Reflecting and Extending

 Using small green triangles, what are the different polygons you can make with an area of 5N units?

SQUARE AREA

GOAL OF THE INVESTIGATION

Students explore measuring area in square inches. They determine the exact and approximate areas of polygons, as well as create some with a measure of 3 square inches.

FACILITATING THE INVESTIGATION

If students are having difficulty visualizing the area of the polygons in square units, they may find it helpful to trace the polygons on square dot paper (page 66).

For Activity 2, students can be encouraged to join more than one of the same type of polygon to help visualize a solution.

Some students may have trouble getting started on Activity 4. It might help them if you ask all the students to approximate the area of the same polygon and then share their different methods.

SAMPLE SOLUTIONS AND DISCUSSION

 A = 4 sq in.   B = 1 sq in.   C = 2 sq in.   D = 4 sq in.   E = 2 sq in.   F = ½ sq in.

Putting 4 Ds together, with the longest side as the edge, can be done to show that D has an area of 4 square inches. The areas of E and F can be found by putting 2 Es (or 2 Fs) together to form a square, covering it with B's and taking half that amount.

 We say that B is 1 square inch because it measures an inch on each side and is shaped like a square.

 Since these are only estimates, some variation in solutions is expected. You may choose to have students record solutions as a range; for example, G is between 3 and 4 square inches. These solutions are given to the nearest ¼ square inch.

G about 3.5 sq in.	H about 2.5 sq in.	I about 1.75 sq in.
J about 1.75 sq in.	K about 1.25 sq in.	L about 1 sq in.
M about 1 sq in.	N about 0.5 sq in.	O about 0.5 sq in.

REFLECTING AND EXTENDING

 Paper is usually 8.5 x 11 inches yielding an area of 93.5 square inches. Area of desks will be about 6 times that. Since it won't be possible to cover the whole surface with small squares (B), students may conceive of the idea of using a ruler to help determine the area, or they may measure the length and width with the B shapes. Either approach should help them begin develop the idea of area as length times width.

Square Area

Starting the Investigation

Most often area is measured in square units.

 How many square inches cover polygons A through C, if B is 1 square inch?

 Explain how to find the areas of polygons D through F in square inches.

 Why do you think B is 1 square inch?

Going Further

If you trace the other Power Polygons on square dot paper you can make an estimate of their area in square inches. As many corners as possible should touch dots when you record them on square dot paper.

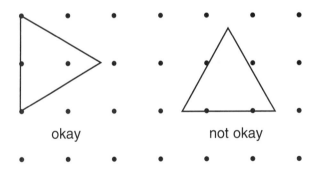

okay   not okay

 About how many square inches are in polygons G through O?

Reflecting and Extending

 a. About how many square inches cover this paper?

b. About how many square inches cover your desk top?

PERIMETER

GOAL OF THE INVESTIGATION

Students first use estimation to identify Power Polygons of equal perimeter. They then identify and use appropriate units of measure to find a good approximation for the perimeter of each of the pieces.

FACILITATING THE INVESTIGATION

Discuss the students' responses to Activity 1 to be sure they can identify a unit of measure for finding perimeter.

Some students may find the perimeter of the quadrilaterals using the edge of the small square (B) since they already know it is 1 inch. However, other students may find this very difficult since many of the sides of the triangles cannot be expressed in whole inches.

If students are having difficulty getting started, you may want to suggest using a ruler. A metric ruler with millimeters would simplify the process of measuring.

SAMPLE SOLUTIONS AND DISCUSSION

 This investigation should help students develop techniques for making estimations. These might include: rotating all the sides of one shape against another, rotating a shape and tracing the length of each side in one connected straight line segment, or using one side of a shape as a unit of length and repeating it around the shape.

Perimeter of: A = G   B = M = O   C = H = I

 Any linear measure could be used to find the distance around a shape. Inches, centimeters, and millimeters, however, are particularly good units of measure for the perimeter.

	A	B	C	G	K	M	O
in.	8	4	6	8	5	4	4
mm	208	104	156	208	130	104	104

 Several of the triangles' sides do not measure in whole-inch or whole-centimeter units. They are best measured with millimeters, since smaller units are more precise.

	D	E	F	I	J	L	N
in.	9 fl	6 ⅞	3 ⅜	6	7 fi	4 fl	3
mm	246	176	88	156	194	128	78

REFLECTING AND EXTENDING

 A + A, G + G, A + G, 4 Cs, 5 Bs, 5 Ms, 5 Os and combinations for Bs, Ms and Os can be attached to yield polygons with perimeters of 12 inches.

Perimeter

Starting the Investigation

Perimeter is different than area. It is about measuring the distance around a shape, not about covering it.

 Which Power Polygons would you guess have equal perimeters?

 Which units of measure could you use to measure the distance around a shape? Which would be best for the Power Polygons? Why?

 What is the perimeter of each different quadrilateral in the set of Power Polygons?

Going Further

 Which unit of measure would be best for measuring the triangles in the set of Power Polygons? Why?

 What is the perimeter of each different triangle in the set of Power Polygons?

Reflecting and Extending

 Which Power Polygons can you connect on a side to make new polygons with a perimeter of 12 inches?

GROWING PERIMETERS

GOAL OF THE INVESTIGATION

Students identify the perimeter patterns that emerge as rectangles are joined on a side. Some students may be able to describe these patterns as algebraic functions.

FACILITATING THE INVESTIGATION

Students may find identifying perimeters easier if they use square dot paper (page 66) to record pictures of polygons A–F and triangular dot paper (page 65) for polygons G–O.

Students may record their findings in a chart like the one below. Some students may be able to identify the algebraic formulas if they understand the concept of using a variable. You might ask them to consider how the number of small squares (B) can be used to evaluate the perimeter without actually making the figure.

SAMPLE SOLUTIONS AND DISCUSSION

 Students may notice the increase of 2 inches with each new piece. If they say it is the even numbers, ask whether it is all the even numbers, so they reflect on the fact that the sequence begins at 4.

Number of Bs	1	2	3	4 . . .	n
Perimeter in inches	4	6	8	10 . . .	$2n + 2$

 Although the square that is added each time has a perimeter of 4 inches, 2 inches are enclosed inside the figure each time. Students might say that each square added 2 inches to the perimeter, but the end pieces each add an additional inch.

 The rhombi, M and O, with side lengths of 1 inch yield the same perimeters.

 Using the larger square, the perimeter grows in a similar fashion, but it increases by 4 inches rather than 2 inches with each additional piece.

Number of As	1	2	3	4 . . .	n
Perimeter in inches	8	12	16	20 . . .	$4n + 4$

 Adjoined on the 2-inch sides

Number of Cs	1	2	3	4 . . .	n
Perimeter in inches	6	8	10	12 . . .	$2n + 4$

Adjoined on the 1-inch sides

Number of Cs	1	2	3	4 . . .	n
Perimeter in inches	6	10	14	18 . . .	$4n + 2$

REFLECTING AND EXTENDING

Number of Bs	1	4	9	16 . . .	n
Perimeter in inches	4	8	12	16 . . .	$4n$

It is unlikely that many students will express the rule as a square root but they may see that there is an increase by 4 each time.

Growing Perimeters

Starting the Investigation

Make bigger and bigger trains of square B, like this:

1 What happens to the perimeter of square B as more Bs are connected?

2 Why does this happen?

3 What other polygon's perimeter would grow the same way?

Going Further

4 What happens to the perimeter of square A as more As are connected to it?

5 What happens to the perimeter of rectangle C as it grows longer by adding on more Cs?

Reflecting and Extending

6　a. Use the small squares to make a 2-by-2 square, a 3-by-3 square, and a 4-by-4 square.

　b. What happens to the perimeter of squares that grow into bigger and bigger squares?

7 AREA AND PERIMETER

GOAL OF THE INVESTIGATION

Students explore the difference between perimeter and area. Students discover that shapes with the same areas do not always have the same perimeter and vice versa.

FACILITATING THE INVESTIGATION

Some students may need to organize their thinking by using a chart. See the sample below. Square dot paper can be found on page 66. Students may start to quickly eliminate some of the polygons since polygons A–F are the only ones that will work

SAMPLE SOLUTIONS AND DISCUSSION

 They must be made from one of the combinations in the chart on the right.

Many of these combinations may result in more than one visually distinct new shape.

B (1 sq")	C (2 sq")	E (2 sq")	F (½ sq")	
1	1			B + C
1		1		B + E
1			4	B + 4F
2			2	2B + 2F
3				3B
	1		2	C + 2F
		1	2	E + 2F
			6	6F

 The whole number perimeters all equal 8 inches, however, other perimeters are possible when fractional values are allowed. The hypotenuse of F measures about 1.4 inches. When this side is exposed in some of the above combinations or when it encloses part of the side of another polygon, several other perimeters are possible.

 The polygons made from 4 Bs all have perimeters of 10 inches, except the square which has a perimeter of 8 inches. Polygons made from other combinations of shapes will have perimeters between 8 and 10 inches. For example, 2B + 4F could be configured with a perimeter of 8⅞ inches. Students don't need to calculate a measurement to see that the perimeter is greater than the square. When traced on dot paper it is visually apparent that the diagonal of the square inch is longer than the side.

 The smallest perimeter is 10 inches and it can be made in different ways.

REFLECTING AND EXTENDING

 Again, the answer is the square.

 Students should notice such things as:
- perimeter and area measure different attributes,
- shapes that look different can sometimes have the same area or the same perimeter, and
- perimeter is the distance around whereas area is the amount of units that covers.

Area and Perimeter

Starting the Investigation

Join the Power Polygons together on a side so that angles touch.
The new polygons must have an area of 3 square inches.

 What are the possibilities? Record the possibilities on square dot paper.

 a. What are the perimeters of these new polygons?

b. What do you notice about the perimeters?

Going Further

 Which shape with an area of 4 square inches has the smallest perimeter?

 Which shape with an area of 5 square inches has the smallest perimeter?

Reflecting and Extending

 Which shape with an area of 9 square inches has the smallest perimeter?

 What have you noticed about the area and perimeter of the polygons you and your classmates made?

Fractions

HALVES

Goal of the Investigation

Students begin investigating fractions by finding the relationships in the Power Polygons set that show halves.

Facilitating the Investigation

Students will need copies of square dot paper (page 66), triangular dot paper (page 65), and the Power Polygons Workmat (page 64).

Usually students are familiar with the concept of ½ by this grade level and will have little difficulty finding some of the solutions. If you have enough polygons, you may want to suggest that students leave each pairing they find visible, so you can check their thinking, and so they can be more efficient in their search.

Students having difficulty may find it easier to build on the Power Polygons Workmat.

For Activity 2, remind students that the A–F polygons trace better on the square dot paper and the G–O polygons trace better on the triangular dot paper.

Sample Solutions and Discussion

C and E are ½ of A	F is ½ of B	B is ½ of C	E is ½ of D
I and J are ½ of G	K is ½ of H	L is ½ of I	L is ½ of J	N is ½ of M

There are some other halves relationships that can be argued logically. For example, B = ½ of E because A = 2E = 4B. These will be difficult for most students to fully grasp at this age and level of experience. However, students who come to this type of thinking on their own should be acknowledged and asked to prove their answers with the polygons.

A, D, G, H, and O are not half of any of the polygons. M is less obvious and some students may not see that M = 2N and I = 4N, therefore M = ½ of I.

Since there are many possible solutions for this section, you may wish to have partners validate each other's work. Essentially, the new polygons should be made by putting together two of one type of polygon. Since polygons can be concave it is not necessary that sides adjoin exactly.

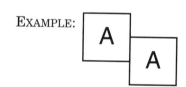

EXAMPLE:

Note: How the two polygons adjoin will be an important part of the description of the new shape. Otherwise students may match descriptions with the wrong shape in the matching game that follows.

Reflecting and Extending

Have partners edit each other's descriptions and put their efforts together to make cards for another pair of students as a matching challenge.

Halves

Starting the Investigation

K is one half of H because two K polygons can cover an H polygon exactly. This can be written like this: K = ½ H.

a. What other statements about one half can you make using the Power Polygons?

b. Which polygons are not half of another polygon in the set?

Going Further

a. Use dot paper to create new polygons so the polygons you listed in Activity 1b will be one half of at least one of the new polygons.

b. Use sentences to describe the attributes of each new polygon you created. Be sure to include a statement that uses ½.

Reflecting and Extending

Make a matching game of the new polygons by putting the shapes and the attributes on separate cards.

This shape has 4 sides.
Opposite sides are equal.
O is fi of this shape.

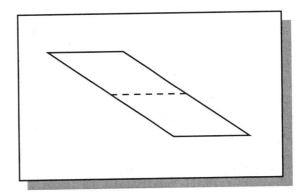

UNIT FRACTIONS

GOAL OF THE INVESTIGATION

Students use the Power Polygons to explore unit fraction relationships other then one half. They extend the concept to shapes they create themselves.

FACILITATING THE INVESTIGATION

For students who have little or no experience with the concept of fractions as part of a whole, you may want to do a paper folding activity prior to this investigation. Have students fold papers in half, fourths, and eighths so they can begin to see the part/whole relationships that embody the area concept of fractions.

Some students may find it easier to build on the Power Polygons Workmat (page 64) than to build on the polygon pieces themselves.

If some students don't seem to know where to begin, suggest that they start with polygon A and see if four of any of the other shapes will cover it exactly. Then they should move to polygon B and so on alphabetically. Other students may use a more spontaneous approach.

Students need to understand that each type of fractional part of a polygon is equal in area to the other fractional parts of the same name. For example, for polygon F to be ⁄ of polygon C, four F polygons must cover polygon C.

SAMPLE SOLUTIONS AND DISCUSSION

 Yes, L is one fourth of G because 4 Ls cover G exactly. The concept of "covering exactly" should be included in the response.

Other responses include:

| B is ¼ of A | F is ¼ of C | F is ¼ of E | N is ¼ of I | M and L are ¼ of G |

 M is ⅓ of H   N is ⅓ of K   N is ⅙ of H   F is ⅛ of A   F is ⅛ of D   N is ⅛ of G

There are some other fractional relationships that can be argued logically, such as N = ¼ of J because G = 2J = 8N. These will be difficult for most students to fully grasp at this age and level of experience. However, students who come to this type of thinking on their own should be acknowledged and asked to prove their answers with the polygons.

 ⅕, ⅐, ⅑, and ⅒ cannot be shown with the Power Polygons.

REFLECTING AND EXTENDING

 There are many possible solutions for this activity. It is both helpful and informative to have students check each other's work. As an alternative you may want to have students capture their ideas on transparent dot paper and show them on an overhead. This helps students see that many questions in mathematics have many solutions.

Unit Fractions

Starting the Investigation

A unit fraction is a fraction whose numerator is 1.

a. Is polygon L one fourth of polygon G? How do you know for sure?

b. Which polygons are one fourth of another polygon?

What other fraction relationships can be represented with the Power Polygons?

Going Further

Which of the unit fractions ½, ⅓, ¼, ⅕, ⅙, ⅐, ⅛, ⅑, and ⅒ can't be shown with the Power Polygons? Why?

Reflecting and Extending

Create new polygons on dot paper that can be used to show the missing unit fractions you listed in Activity 3. Write a fraction relationship statement next to each new polygon.

COMMON FRACTIONS

Goal of the Investigation

Students represent many different fractional parts in this investigation. They also exhibit their understanding of the concept of parts and wholes by identifying the whole when a part is given.

Facilitating the Investigation

Have students use only polygons G–O for this investigation. They will need triangular dot paper (page 65) to trace their solutions.

Introduce the investigation with a demonstration. Using an overhead, show polygon G first with 1M on top, then 2Ms and so on until all fourths relationships are represented. Then have a student draw a picture of each on an overhead transparency of the triangular dot paper.

Inexact representations often lead students to form the misconception that four unequal parts can each be ¼ of a figure. Using dot paper can help students be sure that fractional parts of the same name are drawn so that they are equal in area.

Sample Solutions and Discussion

 The following polygons can be used to show the fractional parts:

½: G and I or J	H and K	I and L	M and N	J and L
⅓: H and M	K and N			
¼: G and M or L	I and N			
⅙: H and N				
⅛: G and N				

 2Ns are ²⁄₆ of H
2Ns are ⅔ of K
2Ns are ²⁄₈ of G
2Ns are ½ of I

Reflecting and Extending

 Have students trade puzzles so they can validate each other's work. Puzzles can be placed in a math activity center to be used throughout the school year.

Common Fractions

Starting the Investigation

Use polygons G through O.

 Show all the common fractions that are listed below. Trace the polygons on triangle dot paper. Color the tracings to show your answers.

$\frac{1}{2}$, $\frac{2}{2}$

$\frac{1}{3}$, $\frac{2}{3}$, $\frac{3}{3}$

$\frac{1}{4}$, $\frac{2}{4}$, $\frac{3}{4}$, $\frac{4}{4}$

$\frac{1}{6}$, $\frac{2}{6}$, $\frac{3}{6}$, $\frac{4}{6}$, $\frac{5}{6}$, $\frac{6}{6}$

$\frac{1}{8}$, $\frac{2}{8}$, $\frac{3}{8}$, $\frac{4}{8}$, $\frac{5}{8}$, $\frac{6}{8}$, $\frac{7}{8}$, $\frac{8}{8}$

Going Further

 Work with a partner. Use the information from Activity 1 to find the polygons described below.

If 2Ns are $\frac{2}{6}$, which polygon is the whole?

If 2Ns are $\frac{2}{3}$, which polygon is the whole?

If 2Ns are $\frac{2}{8}$, which polygon is the whole?

If 2Ns are $\frac{1}{2}$, which polygon is the whole?

Reflecting and Extending

 Make some puzzles of your own. Use descriptions like the ones in Activity 2. Use polygons other than polygon N.

GOAL OF THE INVESTIGATION

Students begin to explore fractional relationships of comparison, equality, and addition at a concrete level. They represent these relationships symbolically.

FACILITATING THE INVESTIGATION

Each student should have triangular dot paper (page 65).

If students are having difficulty getting started, tell them to trace several polygon Gs on their dot paper. Then they should cover each G with a different collection of shapes that fit exactly. As these coverings are removed, have students draw the partition lines on the tracing of G and write the name of the fractional part. Finally, they can write the matching number sentences.

Most students will focus on the coverings with all of the same shape first, then the coverings that use more than one shape. It is unlikely that students will find all possible solutions.

SAMPLE SOLUTIONS AND DISCUSSION

 Some possible solutions are:

$G = \frac{1}{2} + \frac{1}{2}$ (2I or 2 J)

$G = \frac{1}{4} + \frac{1}{4} + \frac{1}{4} + \frac{1}{4}$ (4M or 4L)

$G = \frac{1}{8} + \frac{1}{8} + \frac{1}{8} + \frac{1}{8} + \frac{1}{8} + \frac{1}{8} + \frac{1}{8} + \frac{1}{8}$ (8N)

$G = \frac{1}{2} + \frac{2}{4}$ (J + 2L or I + 2L)

$G = \frac{1}{2} + \frac{4}{8}$ (I + 4N)

$G = \frac{1}{2} + \frac{1}{4} + \frac{2}{8}$ (I + M + 2N)

$G = \frac{1}{4} + \frac{6}{8}$ (M + 6N)

$G = \frac{2}{4} + \frac{4}{8}$ (2M + 4N)

$G = \frac{3}{4} + \frac{2}{8}$ (3M + 2N)

$\frac{1}{2} > \frac{1}{4}$	$\frac{1}{2} > \frac{1}{8}$	$\frac{1}{4} > \frac{1}{8}$	$\frac{2}{4} > \frac{1}{8}, \frac{2}{8}, \frac{3}{8}$
$\frac{3}{4} > \frac{1}{2}, \frac{1}{8}, \frac{2}{8}, \frac{3}{8}, \frac{4}{8}, \frac{5}{8}$	$\frac{3}{8} > \frac{1}{4}$	$\frac{4}{8} > \frac{1}{4}$	$\frac{5}{8} > \frac{1}{4}, \frac{2}{4}, \frac{1}{2}$
$\frac{6}{8} > \frac{1}{4}, \frac{2}{4}, \frac{1}{2}$	$\frac{7}{8} > \frac{1}{4}, \frac{2}{4}, \frac{1}{2}, \frac{3}{4}$		

Students may also include statements with fractions of the same denominator, such as $\frac{2}{4} > \frac{1}{4}$ and statements with fractional names for a whole such as $\frac{4}{4} > \frac{1}{2}$.

REFLECTING AND EXTENDING

$\frac{1}{2} = \frac{2}{4}, \frac{4}{8}$ $\frac{1}{4} = \frac{2}{8}$ $\frac{3}{4} = \frac{6}{8}$ $\frac{2}{2} = \frac{4}{4}, \frac{8}{8}$

Students should fold and shade pieces of paper to show fractional equivalencies.

G Whiz

Starting the Investigation

Polygon G can be represented as a combination of fractional parts.
Here is one example:

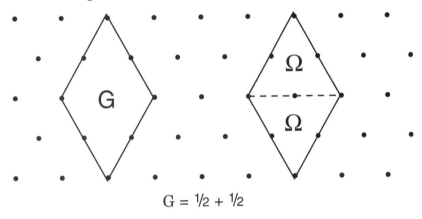

G = ½ + ½

What are some other fractional parts that cover polygon G?

Be sure to trace the polygons as well as write the number sentence.

Going Further

Use the information from Activity 1 to help make decisions about which fractional parts of G are bigger than others.

Record your ideas like this:

½ > ¼ because it takes 4Ms to cover a G but only 2Is.

Reflecting and Extending

a. Use the information from Activity 1 to help you write sentences about fractional parts of G that are equal.

b. Fold paper to prove your ideas about equal fractional parts.

ADDITION

GOAL OF THE INVESTIGATION

Students explore addition of common fractions first with like denominators and then with unlike denominators.

FACILITATING THE INVESTIGATION

Each student will need triangular dot paper (page 65).

The solutions to Activity 1a all relate to the fractional parts of polygon G. Students may need, as a reference, all the exact coverings for G that use the same shape. For example, a covering of polygon G with Ns next to a covering with Ms may help students see that 2 of the 8 parts or ²⁄₈ (2Ns) is equivalent to ¼ (1M). Students may need to leave all the coverings in view during the activities.

Encourage students to estimate the solution for Activity 2a before doing it with the polygons. This can help them reflect on a possible rule for addition that will help when models aren't available.

For Activity 2b, some students may need help seeing that the same type of polygon needs to be used for the covering to express a simple name for the solution. This is a precursor to understanding a need for a common denominator.

SAMPLE SOLUTIONS AND DISCUSSION

¹⁄₈ + ¹⁄₈ = ²⁄₈ or M which is ¼ of G

²⁄₄ + ²⁄₄ = ⁴⁄₄ or G which is 1 whole in this example

¹⁄₈ + ³⁄₈ = ⁴⁄₈ or I which is ½ of G

³⁄₈ + ³⁄₈ = ⁶⁄₈ or 3M which is ¾ of G

²⁄₆ + ²⁄₆ = ⁴⁄₆ or in this case 2M which is ²⁄₃ of H

½ + ⅓ = K + M = 3N + 2N which is ⁵⁄₆ of H

⅓ + ⅙ = M + N or 3N which in this case is ³⁄₆ or K which is ½ of H

REFLECTING AND EXTENDING

Many solutions are possible here. Students are most likely to show problems that already have common denominators. You may want to challenge them to work with denominators that are not the same.

Addition

5

Starting the Investigation

Use polygons G, N, M, and I.

a. Show each of the following addition examples. Record on dot paper.

$\frac{1}{8} + \frac{1}{8}$

$\frac{2}{4} + \frac{2}{4}$

$\frac{1}{8} + \frac{3}{8}$

$\frac{3}{8} + \frac{3}{8}$

b. Give another fractional name that is equal to the answer for each of the examples above.

EXAMPLE: $\frac{1}{8} + \frac{1}{8} = \frac{2}{8}$ which also equals 1M or $\frac{1}{4}$.

Going Further

Use polygons H, K, M, and N.

a. What do you think you would get if you added $\frac{2}{6}$ and $\frac{2}{6}$?

b. What do you think will happen when the fractions below are added?

$\frac{1}{2} + \frac{1}{3}$

$\frac{1}{3} + \frac{1}{6}$

Reflecting and Extending

Make up some addition stories you can show by tracing and shading dot paper. Use two different colors to show the two parts.

ADDITION AND SUBTRACTION

GOAL OF THE INVESTIGATION

Students explore addition of fractions and use the inverse relationship to show subtraction.

FACILITATING THE INVESTIGATION

Students use polygons A–F and square dot paper (page 66).

Students having difficulty may find it easier to find all the different equal shape coverings for polygon A, then B, and so on.

Creating both sentences and drawings should help students see that the same sentence can represent more than one situation. For example, $\frac{1}{2} + \frac{1}{2}$ of polygon A can be shown with Cs or Es.

Writing the subtraction sentences relies on students understanding of fact families. You may want to review this concept with whole numbers before asking students to extend the idea to fractions. They can warm up for Activity 2 by writing fact family stories for numbers 1–7. For example: $1 + 4 = 5$, $4 + 1 = 5$, $5 - 4 = 1$, and $5 - 1 = 4$.

SAMPLE SOLUTIONS AND DISCUSSION

 Polygon A will yield halves (2Cs or 2Es), fourths (4Bs), and eighths (8Fs). Polygon D yields only halves (2Es) or eighths (8Fs).

Students who are organized should easily find all the sentences that can be made with differing amounts of fourths or eighths. Some students may even combine shapes and explore sentences that total more than a whole, i.e. $\frac{5}{8} + \frac{6}{8} = \frac{11}{8}$ or $1\frac{3}{8}$. Be sure to discuss this possibility with the class at some point.

 Students should be encouraged to show sentences with like and unlike denominators.

$\frac{1}{2} + \frac{1}{4} = \frac{3}{4}$	$\frac{3}{4} - \frac{1}{2} = \frac{1}{4}$	$\frac{3}{4} - \frac{1}{4} = \frac{1}{2}$	$\frac{1}{4} + \frac{1}{8} = \frac{3}{8}$
$\frac{3}{8} - \frac{1}{4} = \frac{1}{8}$	$\frac{3}{8} - \frac{1}{8} = \frac{2}{8}$ or $\frac{1}{4}$	$\frac{1}{4} + \frac{2}{8} = \frac{4}{8}$	$\frac{4}{8} - \frac{1}{4} = \frac{2}{8}$
$\frac{4}{8} - \frac{2}{8} = \frac{2}{8}$ or $\frac{1}{4}$	$\frac{1}{2} + \frac{3}{8} = \frac{7}{8}$	$\frac{7}{8} - \frac{1}{2} = \frac{3}{8}$	$\frac{7}{8} - \frac{3}{8} = \frac{4}{8}$ or $\frac{1}{2}$

As you go over possible sentences and their solutions, have students begin to reflect on rules for finding equivalent fractions and sums when denominators are not common. Have them use polygon H to check if the rules they create apply in another situation.

REFLECTING AND EXTENDING

 Making up stories may be difficult for some children. For those students, you can suggest that they stay with a theme for all their stories, such as, parts of a pizza. Other students may just need help seeing the analogy between whole number stories and fraction stories. For example: 5 books take away 3 leaves 2 so $\frac{5}{8}$ of a paper with $\frac{3}{8}$ removed leaves $\frac{2}{8}$ or $\frac{1}{4}$.

Addition and Subtraction

Starting the Investigation

Use polygons A through F.

 Make addition number sentences about the fractional parts of polygon A. Use dot paper to show what you did with the shapes.

> EXAMPLE: Two Bs cover part of polygon A.
> One C covers part of polygon A.

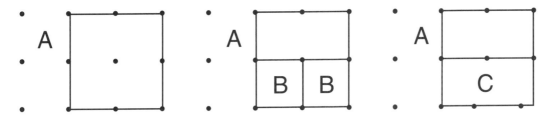

$1/4 + 1/4 = 2/4$ which also equals $1/2$

Going Further

 Subtraction is the opposite of addition. Use the information you found in Activity 1 to write fraction subtraction sentences.

> EXAMPLE: Three Fs cover part of polygon A.
> Two Fs cover one B.

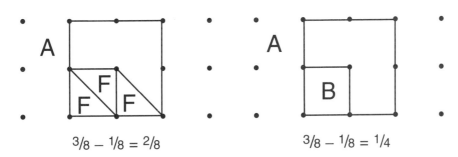

$3/8 - 1/8 = 2/8$ $3/8 - 1/8 = 1/4$

Reflecting and Extending

 Make up some stories for the number sentences you wrote.

> EXAMPLE: I had $3/8$ of a cracker left. I broke off $1/8$ for my brother and then I had only $2/8$ left. This is the same as $1/4$ of the cracker.

FRACTION RIDDLES

GOAL OF THE INVESTIGATION

Students use fraction concepts to solve and create logic riddles.

FACILITATING THE INVESTIGATION

Remind partners that they should check each other's logic.

Some students may find it helpful to have the Power Polygons Workmat (page 64) available as they search for solutions.

There is more than one answer for each question so students will be challenged to think broadly.

When students face more than one clue, they may benefit from the suggestion of selecting the polygons that meet the first clue then discarding elements from that group which do not meet the second clue.

SAMPLE SOLUTIONS AND DISCUSSION

a. B, F, M, L, N

b. A, D, G, H

c. A, D, H, G

d. F, N, O

These questions provide opportunities for students to be creative. Have them do more than one riddle for each solution to think think flexibly.

"I am less than C" would be a simple response for Activity 2a. A more sophisticated solution might be "I am a fractional part of G and I, but not of N."

a. B

b. E, C

c. B, F, N, O

REFLECTING AND EXTENDING

There are many possibilities for student created riddles. Encourage students to use several clues in the riddles they develop. This will insure they challenge themselves as well as the other students who will try them. Having another student try a riddle encourages peer editing rather than you doing the correcting. If you would like parents to see the work you have been doing with students, send home some of the riddles with a copy of the workmat.

Fraction Riddles

Starting the Investigation

Which polygons would make each of the following statements true?

a. I am ¼ of another Power Polygon.

b. I am bigger than ½ of polygon A.

c. I am bigger than ⅚ of polygon H.

d. I am smaller than ⅔ of polygon K.

a. If polygon M is an answer what could the riddle have said?

b. If polygon E is an answer what could the riddle have said?

Going Further

These riddles have more than one clue. Which Power Polygons can they be?

a. I'm bigger than ⅓ of polygon H and smaller than ½ of polygon H.

b. I'm equal to the sum of ⅖ of polygon A and ¼ of polygon A.

c. I'm less than ½ – ⅛ of polygon G.

Reflecting and Extending

Make some riddles of your own. Give them to a classmate to solve.